Ocean Coloring Book

Underwater Coloring Book for Adults containing Seascapes, Fish, Sealife, Coral, Sea Creatures, Marine Life and More

by The Coloring Book People

ISBN-13: 978-1522850038

ISBN-10: 1522850031

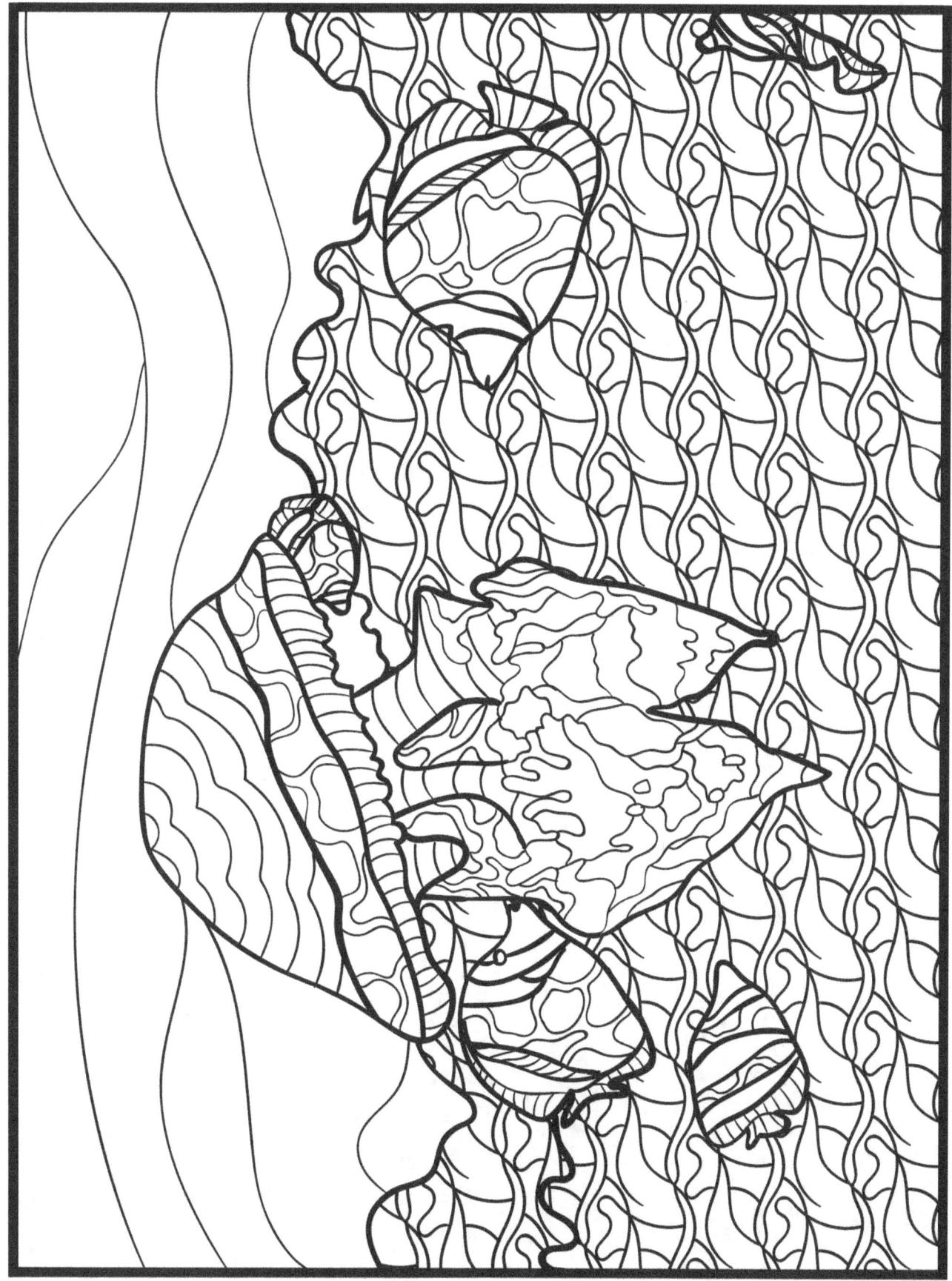

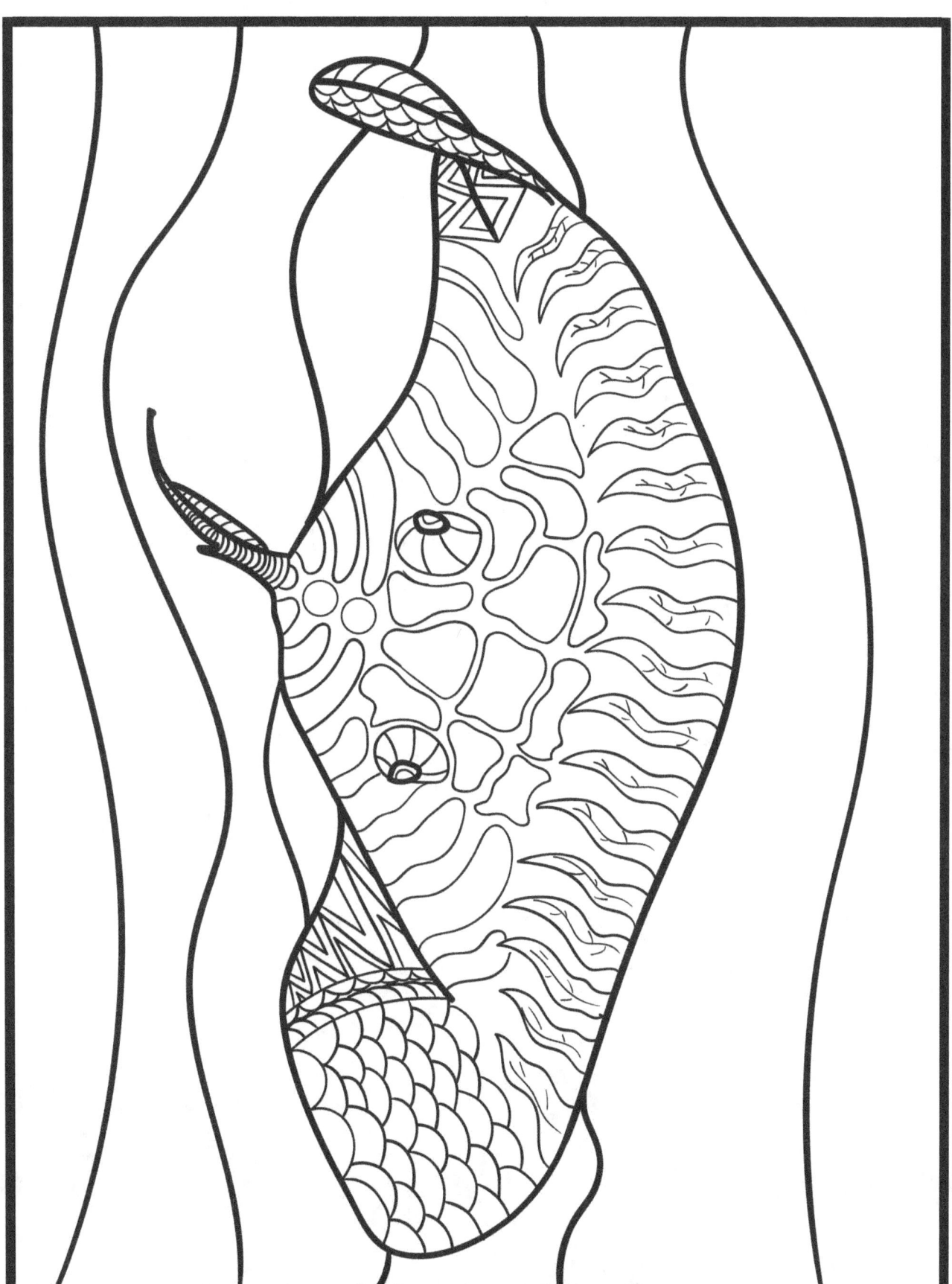

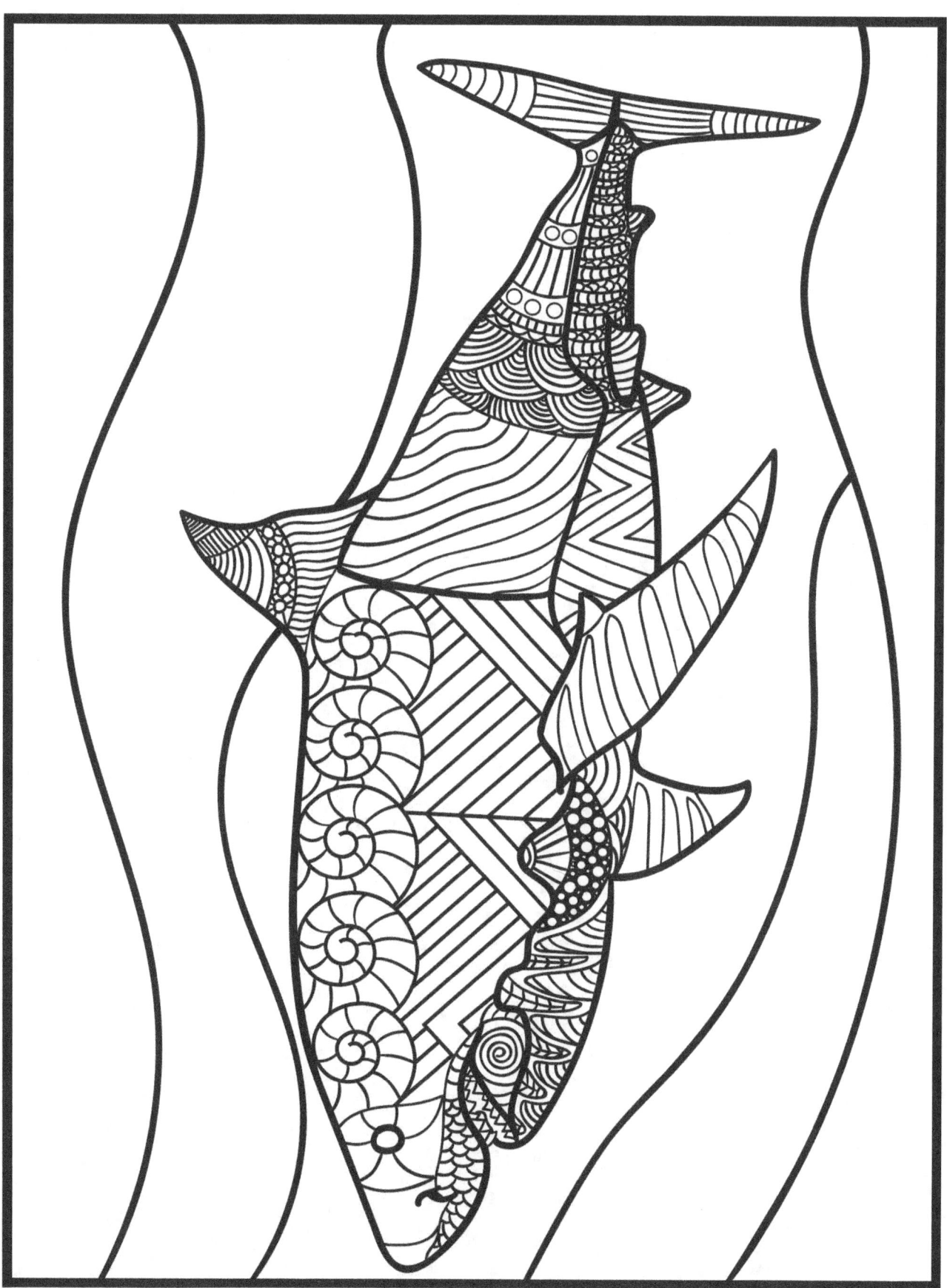

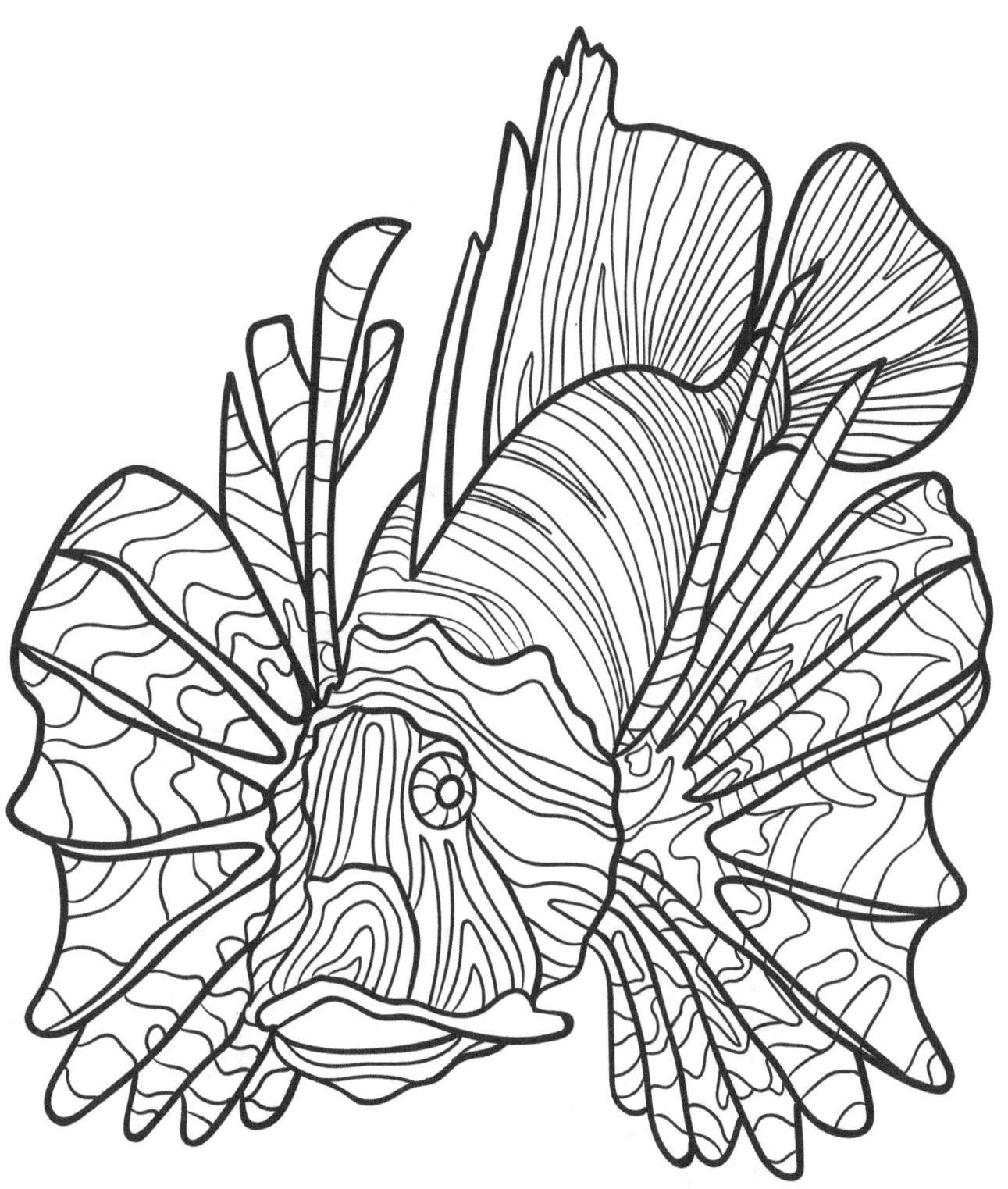

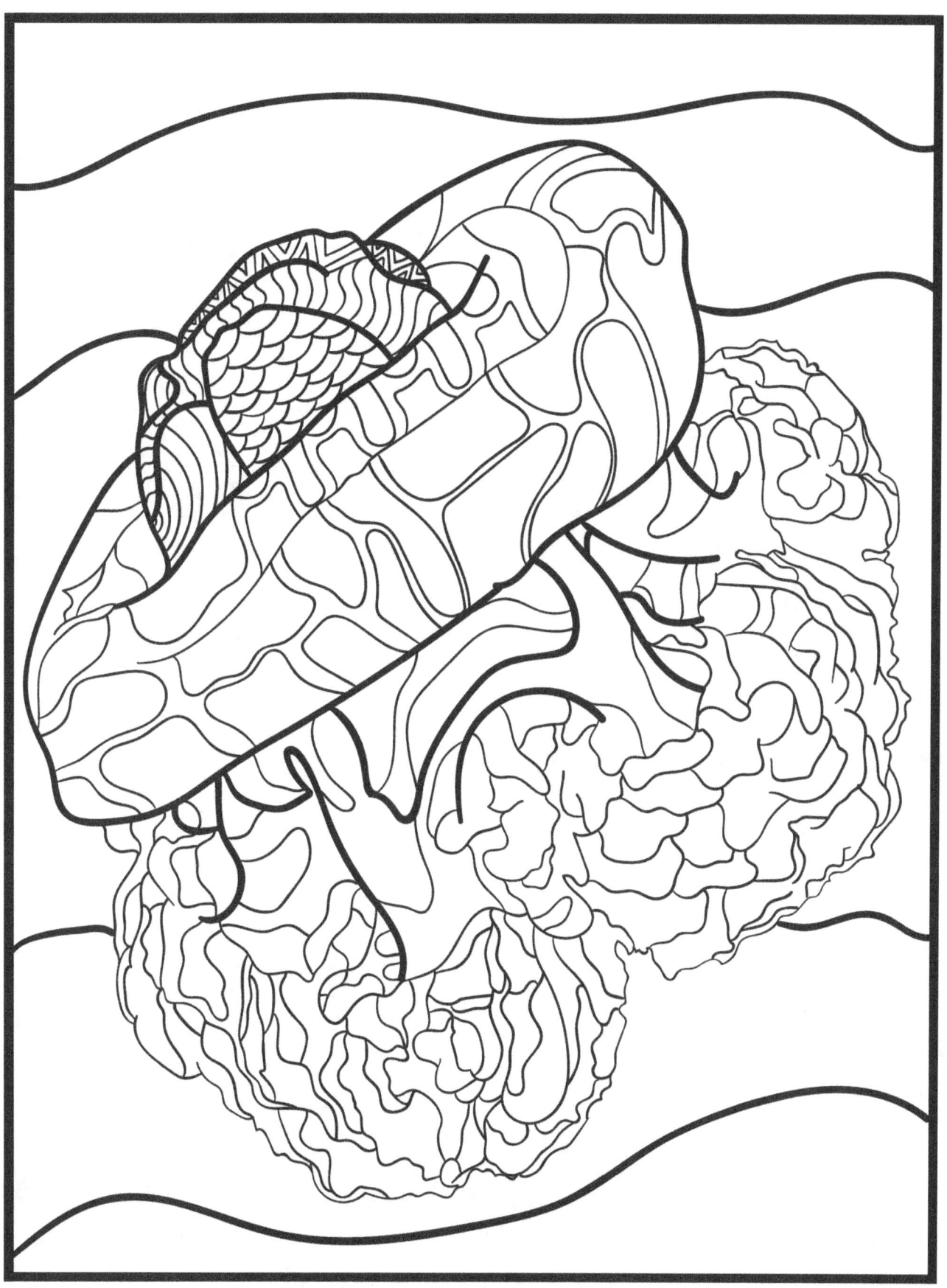

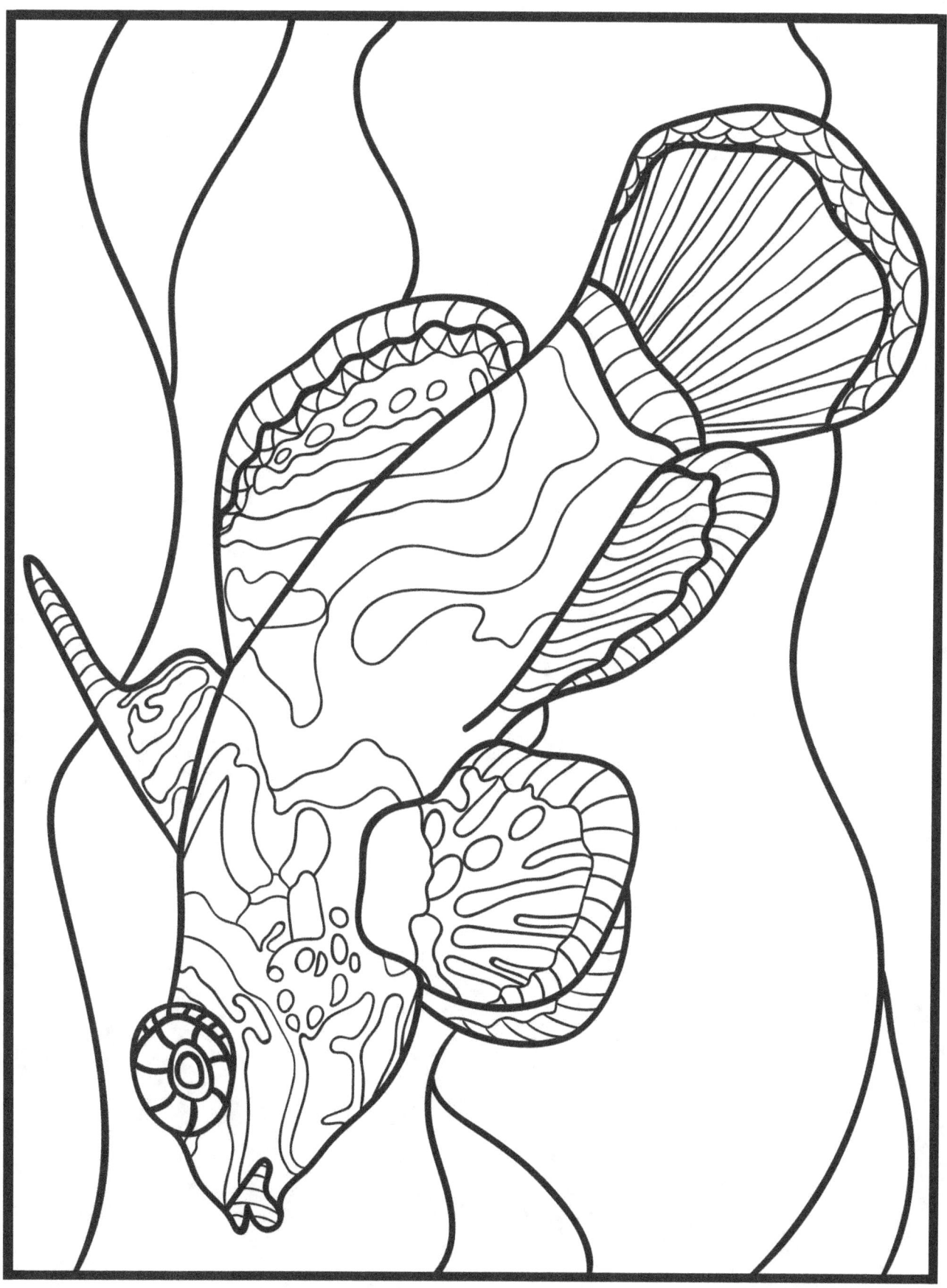

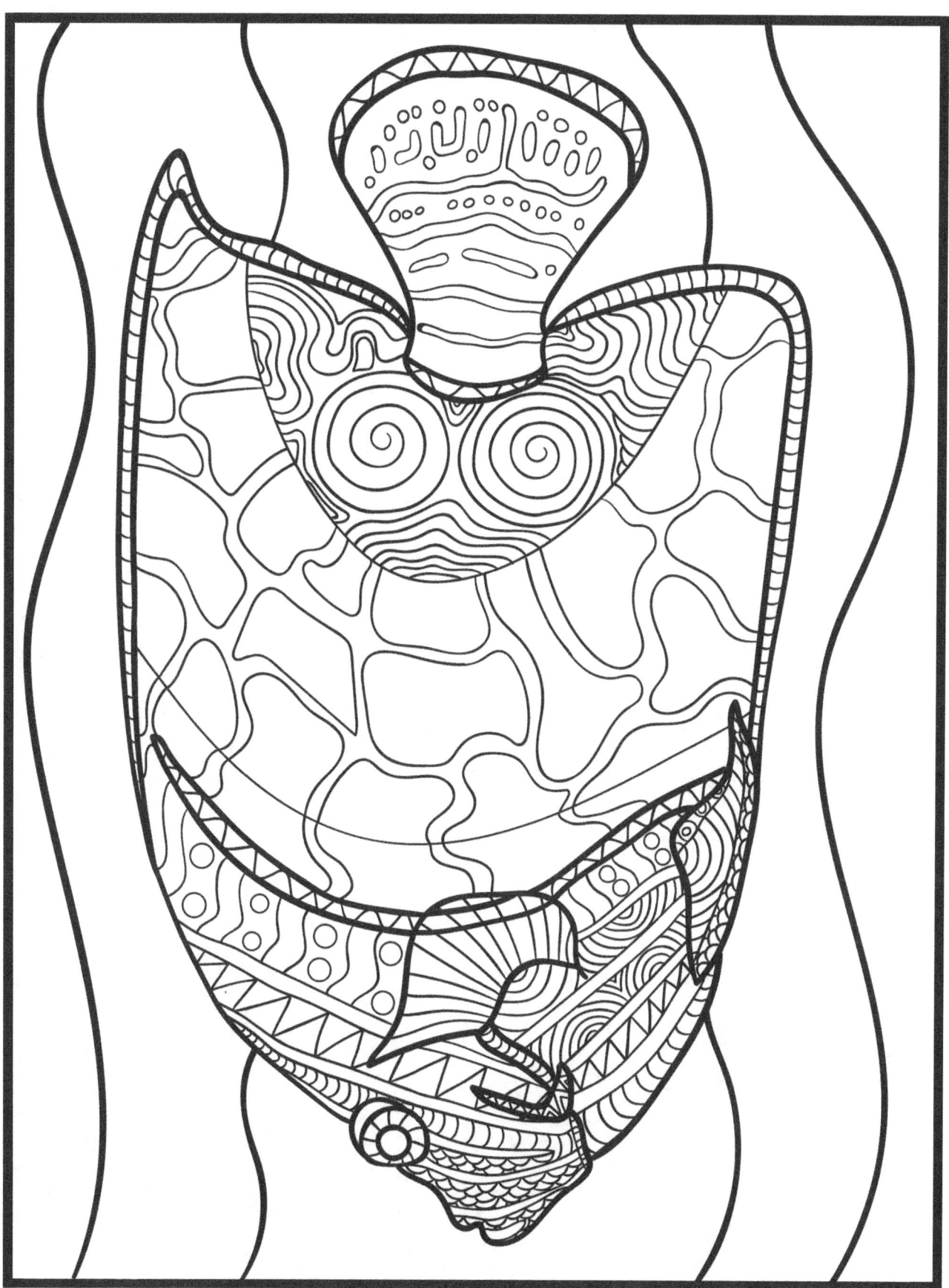

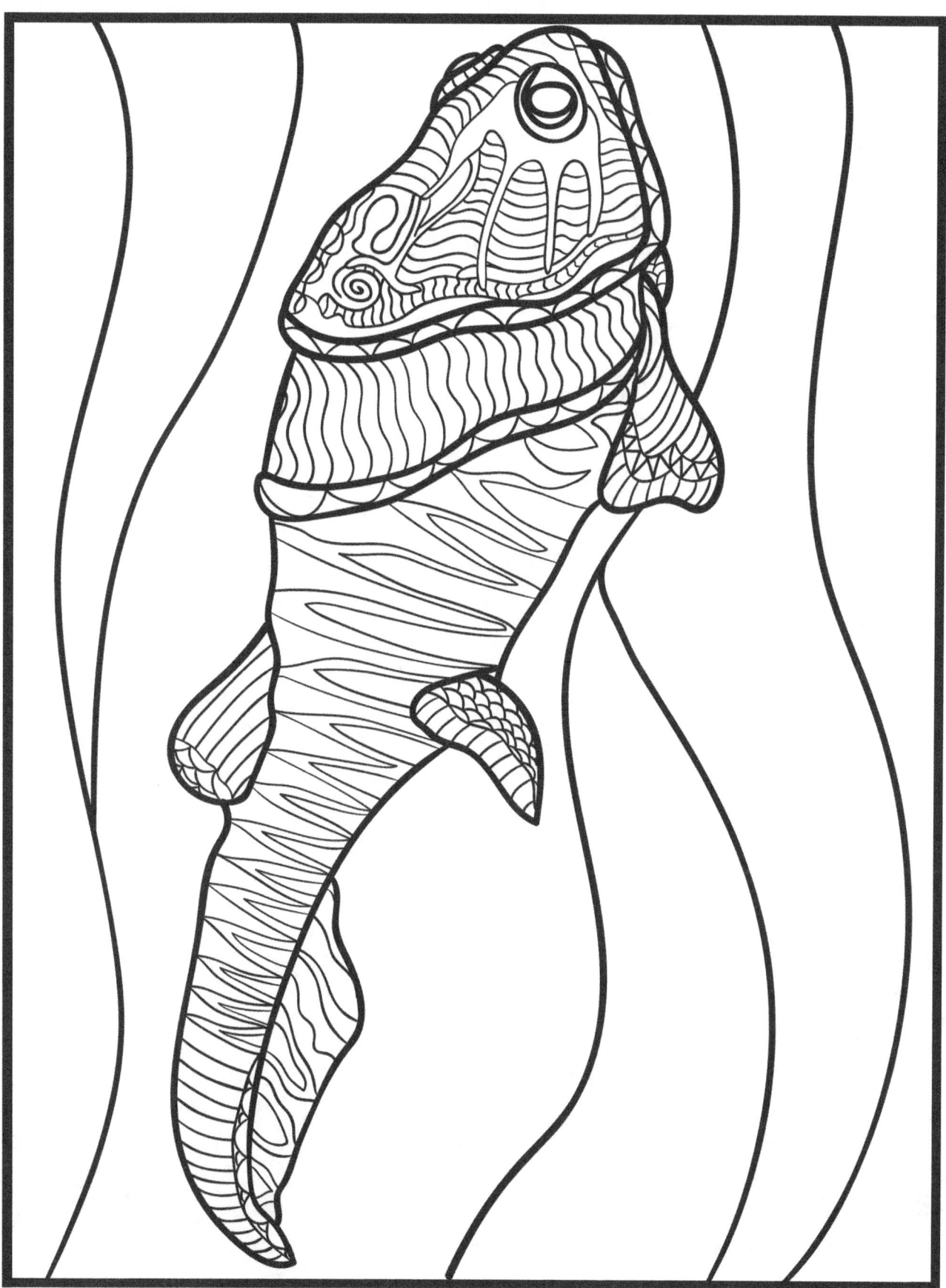

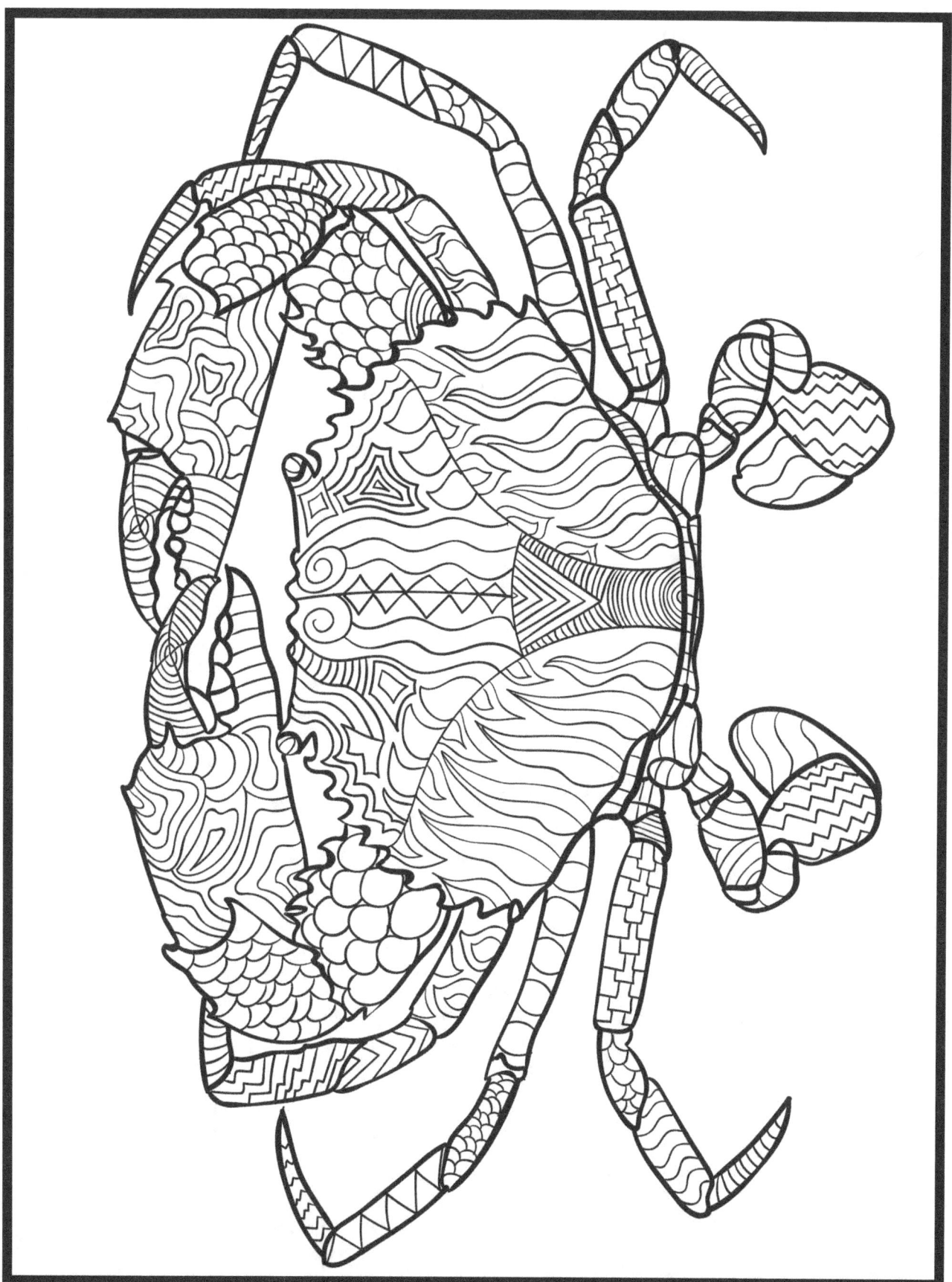

www.ingramcontent.com/pod-product-compliance
Lightning Source LLC
Chambersburg PA
CBHW081404170526

45166CB00010B/3203